Building A Stronger America

A Roadmap to Health, Prosperity, and National Renewal

Jean R. Jean Philippe

Jean Philippe
PRESS

Building A Stronger America

A Roadmap to Health, Prosperity, and
National Renewal

Disclosure on originality and AI usage

While AI tools such as ChatGPT and Perplexity were used during the preparation process to refine clarity and polish language, the core narrative and creative expression remain uniquely the author's. Professional copyediting was also provided by Andrea Gomez.

Printed in the United States of America
First Edition - 2026
Email: jeanphilippe@godrememberscv.com

Epigraph

"Great nations are built by the steady work of ordinary people."

Anonymous.

Acknowledgments

This book was written with deep respect for the millions of ordinary Americans who continue working, building, teaching, serving, creating, and caring for others every day despite growing challenges and uncertainty.

I would like to thank the many thinkers, writers, researchers, educators, healthcare professionals, public servants, business leaders, and community voices whose work on public health, education, housing, infrastructure, agriculture, economics, and national renewal helped shape the ideas explored in these pages.

I am especially grateful to the people who still believe decline is not inevitable—and that practical action, honest leadership, and shared responsibility can help build a stronger future.

This project was also inspired by countless conversations with people from different backgrounds who care deeply about the future of America and want to see the country rise again through unity, resilience, discipline, and opportunity.

Most of all, thank you to the readers.

Your willingness to think seriously about the nation's future is itself a sign of hope. Strong countries are renewed when ordinary people remain willing to engage, reflect, participate, and work toward something greater than themselves.

The future is never guaranteed.

But it is always shaped by those willing to help build it.

Dedication

To the hardworking men and women of America—to the families, workers, students, educators, healthcare professionals, first responders, and community leaders who continue to believe in the promise of a stronger future.

This book is dedicated to all those who refuse to accept decline as destiny and who understand that America's greatest strength has always been its ability to rebuild, renew, and rise through resilience, unity, and hope.

May future generations inherit a nation that is healthier, safer, more prosperous, and more united than ever before.

Table of Contents

Introduction

America at a Crossroads.

America is not collapsing overnight.

Nations rarely do.

Instead, decline happens quietly at first—one struggling family at a time.

A parent forgoes medical care because the cost is too high.

A young graduate starts adult life buried in debt.

A worker drives over cracked roads to a job that no longer seems secure.

A town loses another business, another schoolteacher, another sense of hope.

Little by little, people begin to feel something slipping away.

Not just money.

Not just opportunity.

But confidence in the future itself.

And once a nation loses confidence, everything becomes harder.

People stop believing hard work will pay off. They stop trusting institutions. They cease to believe that tomorrow will be better than today. Cynicism replaces ambition. Survival replaces optimism. Division replaces unity.

But history shows something important:

Decline is not destiny.

Strong nations can be recovered. Broken systems can be repaired. Economies can be rebuilt. Communities can be restored. And countries that lose their footing can rise again—if they are willing to face reality honestly and act with discipline and purpose.

That is what this book is about.

Not blame.

Not ideology.

Not empty slogans.

This book is about rebuilding the foundations that make a nation strong.

Because America's biggest problems are not isolated. They are deeply interconnected.

- When healthcare becomes unaffordable, families fall behind.
- When education costs too much, opportunity shrinks.
- When housing becomes unreachable, stability disappears.
- When addiction spreads, communities weaken.
- When infrastructure crumbles, productivity slows.
- When schools fail, the future suffers.
- When people lose trust in institutions, national unity begins to fracture.

These problems feed one another.

And if they are ignored long enough, even a powerful country begins to weaken from within.

But the opposite is also true.

- When people are healthy, economies grow stronger.
- When education improves, innovation expands.
- When housing becomes affordable, families regain stability.
- When communities become safer, businesses invest again.
- When infrastructure works, opportunity spreads.
- When citizens regain hope, nations regain momentum.

Strength compounds.

This book lays out ten practical areas where America can rebuild its foundation and restore long-term strength: health, education, housing, public safety, addiction recovery, infrastructure, family stability, agriculture, and global competitiveness.

Together, these ideas form something larger:

A roadmap for national renewal.

America does not need perfection to rise again.

It needs competence.

It needs courage.

It needs leadership focused less on outrage and more on results.

And it needs citizens willing to believe that rebuilding is still possible.

Because the future of America will not be decided only by politics.

It will be decided by whether the country can once again do the basic things well:

- educate its children,
- support families,
- protect opportunity,
- reward work,
  and give ordinary people a fair chance to build meaningful lives.

The work of renewal begins now.

Chapter 1

Strengthening Public Health

A nation is only as strong as the people who live in it.

When people are unhealthy, discouraged, addicted, or unable to access care when they need it, the entire country weakens. Public health is not simply a medical issue. It is a foundation of economic productivity, family stability, national resilience, and long-term hope.

Too often, healthcare is treated as a system people enter only after something has already gone wrong. But true public health begins much earlier. It begins with prevention, early treatment, mental health support, affordable access, and communities that help people stay healthy instead of forcing them into crisis.

America already possesses extraordinary medical talent, advanced technology, world-class research institutions, and a culture of innovation. Yet having resources is not the same as using them wisely. A stronger health system must focus less on reacting to emergencies and more on preventing them before they happen.

Prevention saves lives.

Regular checkups, screenings, vaccinations, exercise, healthier diets, and early intervention reduce suffering while lowering long-term costs. A nation that invests in prevention becomes healthier, more productive, and more financially stable over time.

Mental health must also become a central part of the national conversation.

Millions of Americans quietly struggle with anxiety, depression, trauma, loneliness, and stress. Too many suffer in silence until their situation becomes overwhelming. When mental health is ignored, schools suffer, workplaces weaken, families fracture, and communities lose stability. A serious renewal strategy must recognize that mental health is health.

The opioid epidemic has revealed another painful truth: addiction does not destroy only individuals. It tears through families, neighborhoods, schools, hospitals, and entire towns. Recovery requires more than punishment or temporary treatment. It requires compassion, long-term support, prevention, accountability, and systems that help people rebuild their lives with dignity.

Affordability remains one of the greatest barriers to care.

No person should avoid medical treatment because they fear financial ruin. When people postpone care, illnesses become more severe, emergencies become more common, and costs rise for everyone. A healthier system is one that allows people to seek help early—before manageable problems become devastating ones.

Public trust matters too.

People are more likely to follow guidance, seek treatment, and support prevention efforts when they trust the institutions serving them. Trust grows through competence, honesty, accessibility, and clear communication. It weakens when systems are perceived as confusing, distant, or unfair.

There is also a deeper cultural message behind public health.

A society that values health sends a signal that human life matters. It tells workers they are more than labor. It tells families their well-being is worth protecting. It tells children their futures are worth investing in.

That message shapes behavior more than many people realize.

Improving public health does not require perfection. It requires common sense, long-term thinking, and steady commitment. It means expanding preventive care, improving mental health services, lowering costs, supporting healthier lifestyles, and treating addiction with seriousness and humanity.

Because in the end, public health is not merely an expense.

It is an investment in people.

And people are the true strength of any nation. If America wants to prosper again, renewal must begin here.

Notes:

Chapter 1 - Strengthening Public Health

1. Center for Disease Control and Prevention, National Center for Health Statistics, 2024.

2. World Health Organization, World Health Statistics Report, 2024.

Chapter 2

Making Education Affordable Again

Education should be a bridge to opportunity—not a lifetime burden of debt.

For generations, Americans believed education was one of the clearest paths to a better life. Hard work in school was supposed to lead to stability, independence, and upward mobility. But for many young people today, that promise appears increasingly out of reach.

The cost of college has risen dramatically while wages have struggled to keep pace. Students graduate carrying debt before they have had the chance to build up savings, buy homes, start businesses, or raise families. Instead of beginning adulthood with optimism, many begin it with financial anxiety.

That pressure reshapes lives.

Young adults delay marriage. They postpone homeownership. They avoid risk. They work jobs they dislike simply to stay afloat. Over time, this weakens not only individuals, but the economy itself.

A strong nation cannot afford to make opportunities unaffordable.

Education must once again become practical, flexible, and connected to real economic value. Not every student needs the same path.

Some will pursue university degrees. Others will thrive through apprenticeships, technical programs, community colleges, or skilled trades.

All of these paths matter.

America needs engineers and electricians, nurses and welders, scientists and mechanics, teachers and technicians. A healthy economy depends on many forms of talent.

Affordability is not about making education free of responsibility. It is about restoring balance. Students should be able to invest in their future without sacrificing their financial future in the process.

Reducing costs, expanding vocational opportunities, strengthening workforce training, and creating clearer pathways into good careers can restore confidence in the value of education once again.

Because education should expand freedom—not limit it.

And if America wants to rise again, it must make opportunity reachable for the next generation.

Notes:

Chapter 2 - Making Education Affordable Again

1.Organisation for Economic Co-operation and Development (OECD), Education Affordability Data.

2. Pew Research Center report on student debt.

Chapter 3

Improving Housing Affordability

A nation cannot remain strong when ordinary working people can no longer afford a stable place to live.

Housing is more than a roof over someone's head. It is the foundation of family life, financial security, community stability, and personal dignity. It is where children grow up, where workers rest after long days, and where families build their future. When housing becomes too expensive, every part of life becomes harder.

Across America, millions of people are feeling that pressure.

Young adults struggle to move out on their own. Families spend enormous portions of their income on rent or mortgage payments. Workers commute farther because they cannot afford to live near their jobs. Older Americans worry about whether they can remain in the homes and communities they spent years building.

The result is more than financial stress.

It is instability.

When too much income goes toward housing, families have less money for food, healthcare, transportation, education, savings, and emergencies. People become less able to take risks, start businesses, change careers, or invest in their future. A society burdened by unaffordable housing slowly loses mobility, flexibility, and optimism.

Housing affordability is therefore not just a local issue.

It is a national strength issue.

A healthy economy depends on workers being able to live near opportunity. Businesses cannot grow if employees cannot afford nearby housing. Communities cannot remain stable if families are constantly forced to move. Economic growth slows when housing shortages drive up costs and reduce mobility.

One major problem is supply.

In many parts of the country, not enough homes are being built to meet demand. Years of restrictions, delays, rising construction costs, and limited development have created shortages that push prices higher and higher. When supply stays low while demand rises, affordability disappears.

America needs more housing—and more types of housing.

Starter homes.

Affordable apartments.

Middle-income developments.

Housing near transportation and job centers.

A practical strategy must encourage responsible construction while reducing unnecessary barriers that make housing harder and more expensive to build.

This does not mean eliminating standards or sacrificing quality. It means creating systems that allow communities to grow responsibly instead of trapping people in scarcity.

Housing also affects something deeper than economics: human stability.

Children perform better in school when families are stable. Parents experience less stress when housing costs are manageable. Communities become stronger when people can remain long enough to build relationships, trust, and local investment.

Affordable housing helps create stronger neighborhoods because it gives people the ability to put down roots instead of constantly living in uncertainty.

There is also a psychological cost to housing insecurity.

When people believe they can work full-time and still never afford a home, frustration grows. Confidence weakens. The future begins to feel unreachable. Over time, that discouragement spreads far beyond the housing market itself.

A country that makes stable living impossible for large parts of its population cannot remain healthy for long.

America does not need a perfect housing system.

But it does need one that works better than the current one.

That means building more homes, modernizing outdated policies, supporting responsible development, improving infrastructure around growing communities, and making it easier for working families to build stable lives.

Because housing is not simply about buildings.

It is about whether people believe they still have a place in the future of their own country.

And if America wants to rise again, it must become a nation where ordinary people can once again afford to live, work, raise families, and build lasting stability.

Notes:

Chapter 3 - Improving Housing Affordability

1. U.S. Census Bureau, Housing Affordability Statistics.
2. Brookings Institution housing policy analysis.

Chapter 4

Reducing Gun Violence

A nation cannot feel truly strong when its people live with constant fear.

Children should not feel unsafe in classrooms. Parents should not worry every time their child leaves for school. Families should be able to gather in public spaces, attend events, walk through neighborhoods, and live daily life without the shadow of violence hanging over them.

Safety is one of the most basic responsibilities of a stable society.

When people no longer feel safe, trust begins to erode. Communities pull apart. Businesses hesitate to invest. Families move away when they can. Fear changes how people live, where they go, and how they see one another. Over time, that fear weakens the social fabric of the country itself.

Gun violence is not only a criminal justice issue.

It is also a public health issue, a mental health issue, a family issue, and a national confidence issue.

Every shooting creates damage far beyond the immediate victims. Families lose loved ones. Children carry trauma. Communities suffer emotional scars that can last for years. Teachers, first responders, healthcare workers, and neighbors absorb the psychological weight of repeated violence.

The cost is enormous—not only in lives, but in trust, stability, and peace of mind.

Reducing gun violence requires seriousness, honesty, and practical action.

There is no single cause, which means there is no single solution.

Some violence is tied to organized crime and illegal activity. Some grows out of untreated mental illness, social isolation, or personal crisis. Some stems from broken homes, lack of opportunity, or cycles of instability that repeat across generations. A serious response must recognize that violence often grows where systems have already failed people in other ways.

That is why prevention matters.

Communities that invest in strong schools, mentorship programs, mental health services, addiction treatment, and economic opportunity often reduce violence before it begins. Young people who feel connected, supported, and hopeful are less likely to fall into destructive paths.

Intervention matters too.

Many violent incidents involve people already known to be at high risk—individuals trapped in cycles of conflict, instability, trauma, or criminal activity. Early intervention programs, community outreach, crisis support, and focused law enforcement can help stop violence before lives are destroyed.

Enforcement also matters.

Laws must be applied consistently, dangerous individuals must be taken seriously, and communities deserve protection from repeated violence. A society cannot maintain order if people believe rules are optional or consequences are meaningless.

But reducing violence is not only about enforcement.

It is also about rebuilding trust.

In healthy communities, neighbors know one another. Families feel invested in local life. Schools, churches, community organizations, and local leaders work together instead of operating in isolation. Social connection itself becomes a form of protection.

America must also confront the cultural exhaustion that repeated violence creates.

When shootings become routine headlines, people slowly begin to normalize what should never feel normal. That numbness is dangerous. A strong society should never accept fear, trauma, or loss of life as ordinary parts of daily existence.

The goal is not merely fewer headlines.

The goal is stronger communities.

Communities where parents feel safe raising children.

Where businesses can grow confidently.

Where schools focus on learning instead of lockdown drills.

Where people can gather without fear.

Public safety is one of the foundations of national strength.

Without safety, opportunity shrinks. Trust weakens. Communities fracture.

But when people feel secure, society becomes more stable, productive, connected, and hopeful.

If America wants to rise again, it must become a country where safety is not a privilege for a few, but a shared reality for everyone.

Chapter 5

Fighting the Opioid Crisis

The opioid crisis has become one of the deepest wounds in modern American life.

It has taken hundreds of thousands of lives, shattered families, overwhelmed hospitals, strained law enforcement, and left entire communities struggling to recover. What began as a medical and pharmaceutical problem grew into a national emergency that reached cities, suburbs, and small towns alike.

No region has been untouched.

Behind every statistic is a real human story:

- a parent lost too soon,
- a child growing up without stability,
- a worker unable to return to normal life,
- a family carrying grief that never fully disappears.

Addiction destroys more than health.

It damages trust, relationships, productivity, and hope.

Communities hit hard by opioid abuse often experience rising crime, declining workforce participation, growing homelessness, and increased pressure on schools, hospitals, and local governments. Businesses struggle to hire dependable workers. Families break under emotional and financial strain. Entire neighborhoods begin to lose stability.

A nation cannot remain strong while so many of its people are being quietly consumed by addiction.

The opioid crisis must therefore be treated not as a side issue, but as a national priority.

That begins with understanding a difficult truth:

Addiction is not solved by punishment alone.

While accountability matters, addiction is also a medical, psychological, and social crisis. Many people battling opioids are struggling with pain, trauma, depression, isolation, hopelessness, or untreated mental health conditions. Others became addicted after legal prescriptions and gradually lost control over their lives.

A serious response must combine prevention, treatment, recovery support, enforcement, and long-term rebuilding.

Treatment must become easier to access.

Too many people who want help face long waitlists, high costs, insurance barriers, or limited treatment options. Some give up before recovery even begins. A nation serious about renewal cannot allow people in crisis to navigate a broken system alone.

Recovery also requires more than detox programs or temporary treatment centers.

People rebuilding their lives often need counseling, job opportunities, stable housing, peer support, and long-term guidance. Recovery is not a single moment. It is a process that takes time, discipline, support, and hope.

Prevention is equally important.

America must reduce the conditions that allow addiction to spread so easily. That includes responsible prescribing practices, stronger education about substance abuse, better mental health care, and earlier intervention for people at risk.

But the opioid epidemic also reveals something larger about the country itself.

Addiction often grows fast where people feel disconnected, discouraged, isolated, or forgotten. Communities struggling with unemployment, family breakdown, declining opportunity, and social instability become more vulnerable to despair—and despair creates openings for addiction to take hold.

This means the fight against opioids is not only about drugs.

It is also about rebuilding human connections, restoring purpose, strengthening families, creating opportunity, and helping people believe their lives still matter.

Communities affected by addiction do not need judgment from a distance.

They need compassion combined with action.

They need leaders willing to invest in recovery instead of simply managing decline. They need systems that help people stand back up instead of writing them off permanently.

Most importantly, they need hope.

Because recovery becomes possible when people believe the future can still be different from the past.

America has overcome great crises before. It can overcome this one too.

But it will require honesty, persistence, and a refusal to treat addiction as someone else's problem.

The opioid crisis is ultimately a fight for lives, families, communities, and the future strength of the nation itself.

And if America wants to rise again, it must help its people rise again first.

Chapter 6

Raising Education Quality

A strong nation must educate its people well.

Not simply enough to pass tests.

Not simply enough to earn diplomas.

But well enough to prepare them for life, work, citizenship, and the challenges of a rapidly changing world.

Education is one of the most powerful forces in any society. It shapes the workforce, strengthens communities, drives innovation, and determines whether future generations are equipped to lead or left struggling to catch up.

When education weakens, nations weaken with it.

Access to education matters, but access alone is not enough. Students also need quality. They need schools that teach them how to think critically, solve problems, communicate clearly, and adapt to change. They need strong reading skills, mathematical understanding, scientific literacy, discipline, and the ability to separate truth from misinformation.

These are not optional skills in the modern world.

They are survival skills.

A country that fails to educate its children properly eventually pays the price through lower productivity, weaker innovation, declining competitiveness, and growing social instability.

Education quality begins with high expectations.

Students rise when standards are clear, serious, and consistent. Lowering expectations may temporarily make systems look better on paper, but it does not prepare young people for reality. True fairness does not mean lowering the bar. It means giving every student the support needed to reach it.

Teachers play a central role in this mission.

No education system can rise above the quality of the people standing in front of the classroom each day. Great teachers do far more than deliver information. They build confidence, discipline, curiosity, and character. They help students discover what they are capable of becoming.

Yet too many teachers today are exhausted, under-supported, and overwhelmed by growing demands.

If America wants stronger schools, it must support stronger teaching.

That means better training, better resources, safer classrooms, reasonable workloads, professional respect, and systems that allow teachers to focus on education instead of constant bureaucracy.

Parents matter just as much.

Children do not learn only at school. They learn at home through routines, conversations, expectations, and daily habits. Families shape attitudes toward discipline, responsibility, effort, and learning long before students enter a classroom.

The strongest education systems are partnerships.

Teachers, parents, schools, and communities must work together instead of operating against one another. When children receive consistent expectations both at home and at school, learning becomes more stable and effective.

Education must also prepare students for the future economy.

The modern world rewards adaptability, technical skills, creativity, communication, and lifelong learning. Students should leave school not only with knowledge, but with the ability to continue learning as industries and technologies evolve.

This includes stronger vocational pathways as well.

Not every student will follow the same route, and that is healthy for the country. Skilled trades, technical careers, apprenticeships, engineering, healthcare, entrepreneurship, manufacturing, and digital industries all require capable, well-trained people. A successful education system respects multiple forms of talent and multiple pathways to success.

Technology presents both opportunities and risks in education.

Digital tools can expand learning, improve access, and connect students to information more quickly than ever before. But technology cannot replace human guidance, discipline, mentorship, and real understanding. Schools must teach students not only how to consume information, but how to think independently about it.

Education is ultimately about more than careers.

It is about citizenship.

A healthy democracy requires people who can reason carefully, communicate respectfully, understand history, analyze information, and participate thoughtfully in public life. Without those abilities, division deepens and trust weakens.

A society that educates well becomes more innovative, more stable, and more confident about its future.

It produces citizens capable of building businesses, solving problems, strengthening communities, and leading responsibly. It creates people who can compete globally while remaining grounded locally.

If America wants to rise again, it must once again become a country that takes education seriously—not only in spending, but in standards, discipline, expectations, and long-term commitment.

Because education does more than shape individual lives.

It shapes the future strength of the nation itself.

Chapter 7

Rebuilding Infrastructure

No nation can remain strong while its foundations are falling apart.

Roads, bridges, ports, rail systems, airports, water lines, power grids, broadband networks, and public transit systems are more than construction projects. They are the physical framework that allows a country to function. They connect workers to jobs, businesses to markets, students to schools, and communities to opportunity.

When infrastructure works well, people often barely notice it.

When it fails, everyone feels the consequences.

A bridge closure can delay thousands of workers. Aging water systems can threaten public health. Traffic congestion wastes time, fuel, and productivity. Weak power grids leave communities vulnerable during extreme weather. Poor internet access limits education, healthcare, and economic opportunity, especially in rural areas.

Over time, these failures create frustration, inefficiency, and declining confidence in the country's ability to maintain even its most basic systems.

Infrastructure may not always dominate political headlines, but it quietly shapes daily life more than almost anything else.

Strong infrastructure creates economic strength.

Businesses depend on reliable transportation and energy systems. Manufacturers need supply chains that move efficiently. Workers need safe roads and dependable transit. Farmers need transportation networks to move products across the country and around the world.

When infrastructure weakens, costs rise everywhere.

Deliveries slow down. Repairs become more expensive. Productivity declines. Economic growth becomes harder to sustain. A nation that neglects its infrastructure slowly becomes less competitive, less efficient, and less prepared for future challenges.

Infrastructure is also about national resilience.

Modern societies face increasing pressure from extreme weather, cyber threats, population growth, and rapidly changing technology. Systems built decades ago are now being pushed far beyond what they were designed to handle. Without modernization, vulnerabilities grow larger every year.

That is why rebuilding infrastructure requires long-term thinking.

A serious nation does not wait for disaster before taking action. It plans ahead. It maintains critical systems before they collapse. It invests in durable projects that strengthen the country for future generations rather than only solving immediate political problems.

Broadband access now belongs in the same conversation as roads and electricity.

In the modern economy, internet access is essential infrastructure. Students rely on it for education. Businesses rely on it for commerce. Hospitals rely on it for healthcare services. Workers rely on it for communication and employment opportunities. Communities left behind digitally are increasingly left behind economically as well.

Infrastructure also reflects national discipline.

A country that maintains its bridges, roads, water systems, airports, and energy networks sends a message that it still believes in building for the future. It shows competence, seriousness, and confidence. It demonstrates that long-term stability matters more than temporary convenience.

There is also a deeper psychological effect.

People notice when the basics stop working. They notice crumbling roads, unreliable transit, failing public systems, and constant delays. Over time, those daily frustrations shape how citizens feel about their country itself. Confidence weakens when people believe decline has become normal.

But the opposite is also true.

Modern, reliable infrastructure creates momentum. It encourages investment, supports innovation, lowers costs, creates jobs, and improves quality of life. It signals that the country is still capable of solving problems and building ambitious things.

Infrastructure projects also create something increasingly rare in modern society: visible progress.

People can see a repaired bridge. They can drive on smoother roads. They can experience faster internet, cleaner water, and shorter commutes. Tangible improvements restore trust because they remind citizens that effective government and national progress are still possible.

Rebuilding infrastructure is not simply about concrete, steel, and cables.

It is about rebuilding confidence in the country's ability to function well.

If America wants to rise again, it must once again become a nation that builds, maintains, modernizes, and plans for the future with seriousness and pride.

Because strong nations are built on strong foundations.

And foundations must be maintained before they fail.

Notes:

Chapter 7 - Rebuilding Infrastructure

1. American Society of Civil Engineers, Infrastructure Report Card, 2025.

Chapter 8

Expanding Paid Leave and Work Stability

A strong nation depends on stable workers, stable families, and stable communities.

But for millions of Americans, modern work has become increasingly unstable. Many people live under constant pressure—worried about missing a paycheck, losing hours, falling behind on bills, or being unable to handle even a small emergency. One illness, one family crisis, or one unexpected disruption can push an otherwise hardworking person into financial hardship.

That kind of instability slowly weakens a country from within.

Work should provide more than income alone. It should provide dignity, structure, purpose, and the ability to build a stable life. When workers constantly feel overwhelmed, exhausted, or financially insecure, families suffer, productivity declines, and communities become more fragile.

Paid leave and work stability may sound like technical policy issues, but they directly affect daily life for millions of people.

A worker without paid sick time may go to work while seriously ill because they cannot afford to stay home. A parent without family leave may be forced to choose between caring for a newborn child and keeping a job. Someone caring for an aging parent may face impossible decisions between responsibility and survival.

No healthy society should force people into those choices repeatedly.

Paid leave is not about avoiding work.

It is about creating a system where people can continue working without collapsing under constant pressure.

Workers who have reasonable flexibility and support are often more productive, more loyal, and more dependable over time. Businesses benefit when employees are healthier, less stressed, and more stable. Lower turnover, stronger morale, and better long-term performance help employers as much as workers themselves.

In that sense, work stability is not only compassionate.

It is economically practical.

The modern economy has also become more unpredictable for many people. Temporary contracts, irregular schedules, unstable hours, and rising living costs create a sense of permanent uncertainty. Even full-time workers often feel like they are barely staying ahead.

That insecurity affects more than finances.

It affects mental health, relationships, parenting, and long-term planning. People become hesitant to start families, buy homes, pursue education, or invest in their future when they feel one setback away from crisis.

A country cannot build long-term strength on widespread exhaustion and instability.

Family life especially depends on balance.

Children benefit when parents have time to recover from illness, attend important events, and remain emotionally present. Strong families create stronger communities, and stronger communities create more stable societies.

Work should support family life—not constantly undermine it.

This does not mean eliminating responsibility or reducing standards.

Hard work will always matter.

But sustainable work systems recognize a basic truth: people are not machines. Human beings occasionally get sick, face emergencies, care for loved ones, and need time to recover physically and mentally. A society that ignores those realities creates unnecessary strain that eventually harms everyone.

Technology and economic change are also reshaping the future of work.

Automation, artificial intelligence, remote work, and global competition are transforming industries faster than many workers can adapt. This makes stability even more important. Workers need opportunities for retraining, skill development, and predictable support during periods of transition.

A resilient workforce is one that can adapt without falling apart.

Countries that support stable work environments often experience stronger productivity, healthier populations, and greater social trust. People perform better when they feel secure enough to focus on long-term goals instead of constant survival.

At its best, work gives people more than wages.

It gives them identity, contribution, purpose, and pride.

But that becomes harder when instability dominates everyday life.

If America wants to rise again, it must create an economy where hard work once again leads to stability rather than permanent anxiety.

Because national strength depends not only on markets and profits.

It depends on whether ordinary people can build steady, dignified lives through their work.

Chapter 9

Increasing Agricultural Production

A strong nation must be able to feed itself.

Food security is not a secondary issue. It is one of the foundations of national stability, economic resilience, and long-term independence. A country that cannot reliably produce and distribute enough food places itself in a vulnerable position—economically, politically, and strategically.

For much of its history, America's agricultural strength has been one of its greatest advantages.

American farmers have helped feed not only the nation, but much of the world. Through innovation, efficiency, hard work, and technological advancement, the United States became an agricultural powerhouse capable of producing enormous quantities of food while supporting millions of jobs and businesses connected to farming.

But maintaining that strength requires constant attention.

Agriculture today faces growing pressures:

- rising production costs,
- supply chain disruptions,
- extreme weather,
- water shortages,
- soil depletion,

- labor shortages,
- global competition,
  and increasing uncertainty in international markets.

If these pressures are ignored, food prices rise, rural communities weaken, and national resilience declines.

Agriculture matters far beyond the farm itself.

A strong agricultural sector supports transportation networks, manufacturing, food processing, equipment industries, energy systems, exports, and local economies across the country. Millions of Americans depend directly or indirectly on the success of agriculture.

When farming struggles, the effects spread far beyond rural America.

Increasing agricultural production is not simply about growing more food at any cost.

It is about producing food more efficiently, more sustainably, and more reliably.

Modern technology has already transformed farming through advanced machinery, precision agriculture, improved irrigation systems, data analysis, biotechnology, and more efficient crop management. Continued investment in research and innovation can help farmers increase yields while conserving resources and reducing waste.

The future of agriculture will depend heavily on adaptation.

Climate conditions are changing. Weather patterns are becoming less predictable. Water resources in some regions are under increasing stress. A strong nation must prepare its agricultural systems for long-term resilience rather than assuming current conditions will remain stable forever.

This means investing in smarter farming methods, improved infrastructure, drought-resistant crops, water management systems, and agricultural research that helps farmers remain productive under changing conditions.

Food affordability also matters deeply.

When food prices rise too quickly, lower-income families suffer first and most severely. Inflation in food costs creates pressure throughout the economy and increases financial strain on millions of households. A productive agricultural system helps stabilize prices and protect families from unnecessary hardship.

Agriculture is also tied to national independence.

Countries that become overly dependent on foreign food supplies expose themselves to major risks during wars, trade disruptions, natural disasters, or global instability. Strong domestic production gives nations greater security and flexibility during uncertain times.

Rural communities deserve attention as well.

Many farming regions across America face population decline, aging infrastructure, economic challenges, and shrinking opportunities for younger generations. If agriculture is to remain strong, rural America must remain livable and economically viable. Farmers, ranchers, and agricultural workers should not feel forgotten by the broader economy they help sustain.

There is also a cultural lesson in agriculture.

Farming teaches patience, discipline, resilience, and long-term thinking. Farmers invest months of work before seeing results. They adapt constantly to forces beyond their control while continuing to provide one of society's most essential needs.

That mindset matters for national renewal.

Strong countries think long term. They build systems designed to endure rather than simply chasing short-term convenience.

Agriculture reflects that principle clearly.

A nation that can feed itself, support its farmers, maintain stable food supplies, and continue innovating in agricultural production is a nation that possesses real resilience.

America already has enormous agricultural strengths:

- vast farmland,
- advanced technology,
- skilled producers,
- research institutions,
- transportation networks,
- and generations of farming expertise.

The challenge is not whether the country can remain agriculturally strong.

The challenge is whether it is willing to invest seriously in protecting and modernizing one of the foundations of its long-term stability.

To strengthen food security, the United States must increase agricultural production while helping farmers remain competitive in global markets and resilient against uncertainty caused by wars and other international crises.

Because food security is not only about survival.

It is about national strength, independence, and the ability to endure uncertainty with confidence.

Chapter 10

Staying Globally Competitive

A nation does not remain strong by standing still.

In a rapidly changing world, countries either adapt and compete—or slowly fall behind. Economic power, technological leadership, innovation, and global influence are never guaranteed permanently. They must be earned and renewed generation after generation.

For decades, America's strength came from its ability to innovate, build, attract talent, and create opportunity on a scale few nations could match. The country became a global leader not simply because of its size, but because it rewarded ambition, encouraged invention, and created an environment where new ideas could grow into world-changing industries.

That competitive advantage cannot be taken for granted.

Today, the global economy is moving faster than ever. New technologies are reshaping industries almost overnight. Artificial intelligence, advanced manufacturing, biotechnology, cybersecurity, robotics, and energy innovation are transforming how nations compete. Countries around the world are investing aggressively in education, infrastructure, research, and industrial development because they understand what is at stake.

The future will belong to nations that continue improving.

And decline often begins when countries assume past success alone will guarantee future strength.

Global competitiveness is not just about economics.

It is about whether a country remains a place where people want to build businesses, invest capital, conduct research, create new technologies, and raise families with confidence about the future.

If innovation slows, opportunity shrinks.

If businesses face constant instability, investment moves elsewhere.

If talented workers and entrepreneurs stop believing they can succeed, national momentum begins to fade.

That is why competitiveness depends on many systems working together.

Strong education creates skilled workers and innovators.

Strong infrastructure supports efficient business activity.

Strong public health keeps the workforce productive.

Affordable housing helps attract and retain talent.

Stable energy systems support industry.

Reliable institutions encourage investment and trust.

Competitiveness is not built by one policy alone.

It is built through national competence across many areas at once.

Innovation remains one of America's greatest strengths.

The country still leads the world in many fields of technology, medicine, entrepreneurship, entertainment, finance, and scientific research. American universities, companies, researchers, and entrepreneurs continue producing breakthroughs that shape the global economy.

But leadership can weaken if complacency replaces ambition.

The countries most likely to succeed in the future will be those willing to invest heavily in research, workforce development, advanced manufacturing, and emerging technologies. They will create systems that reward creativity while maintaining economic stability and national resilience.

America must remain open to talent and ideas.

Throughout history, many of the nation's greatest achievements came from people who arrived with little except determination, skill, and ambition. Attracting capable people from around the world has long been one of America's competitive advantages. A country that welcomes innovation and rewards hard work strengthens itself continuously.

Competitiveness also depends on confidence.

When citizens believe their country is capable of solving problems and adapting to change, they become more willing to invest, create, and take risks. But when pessimism dominates public life, people pull back. Innovation slows. Ambition fades. Fear replaces forward thinking.

National confidence matters economically as much as psychologically.

At the same time, global competitiveness should not be confused with arrogance or endless expansion.

The goal is not domination for its own sake.

The goal is effectiveness.

A competitive nation creates conditions where businesses can grow, workers can succeed, industries can innovate, and communities can remain stable through changing economic conditions. It balances ambition with resilience.

America must also prepare workers for economic transformation.

Automation and artificial intelligence will likely reshape millions of jobs in the coming decades. Some industries will expand dramatically while others shrink or disappear. Workers need education, retraining opportunities, and support systems that help them adapt rather than become permanently displaced.

The countries that handle this transition best will emerge stronger.

Those that fail to adapt may experience deeper instability and division.

Competitiveness is ultimately about more than wealth.

It is about whether a nation still believes in progress.

Strong countries continue building. They continue learning. They continue improving their systems instead of assuming greatness will maintain itself automatically.

America still possesses extraordinary advantages:

- world-class universities,
- deep capital markets,
- innovative businesses,
- natural resources,
- global influence,
- entrepreneurial culture,
- and generations of accumulated knowledge and experience.

The challenge is whether those strengths will continue to be developed seriously and responsibly.

If America wants to rise again, it must once again become a country focused on excellence, innovation, productivity, and long-term thinking.

Because in a competitive world, nations that stop improving rarely remain strong for long.

Chapter 11

Building a Stronger America

America's future is not fixed.

Decline is not inevitable.

Renewal is not impossible.

And the choices made today will shape the kind of nation future generations inherit tomorrow.

For much of its history, America's strength did not come from perfection. It came from resilience—the ability to adapt, rebuild, innovate, and move forward even during periods of crisis. The country faced wars, economic collapse, social unrest, political division, and global competition, yet repeatedly found ways to recover because people believed improvement was still possible.

That belief matters.

A nation becomes stronger when its people still believe the future can be better than the present.

This book has argued that national renewal does not begin with slogans or temporary political victories. It begins with strengthening the foundations that hold a country together:

- healthy people,
- strong families,
- safe communities,
- quality education,
- reliable infrastructure,
- stable work,
- productive agriculture,
- and a competitive economy capable of adapting to change.

These are not isolated issues.

They are connected.

A healthier population creates a stronger workforce.

Better schools produce more capable citizens and innovators.

Affordable housing supports family stability.

Reliable infrastructure strengthens productivity.

Safer communities encourage investment and opportunity.

Strong agriculture protects food security.

Stable work environments reduce stress and improve long-term growth.

Each area reinforces the others.

And when enough of these systems weaken at the same time, decline spreads quietly across society.

But when these systems improve together, strength compounds.

That is how nations rebuild.

Not all at once.

Not through one election or one speech.

But through steady, disciplined improvement over time.

A stronger America will require honesty.

The country cannot solve problems it refuses to acknowledge. Pretending everything is fine when people are struggling only deepens frustration and distrust. Real progress begins when leaders and citizens alike are willing to confront weaknesses directly instead of hiding behind political talking points or empty optimism.

At the same time, honesty should not become hopelessness.

America still possesses enormous strengths:

- a large and innovative economy,
- world-class universities,
- deep entrepreneurial culture,
- scientific leadership,
- natural resources,
- global influence,

and millions of hardworking people who continue building, creating, teaching, raising families, and contributing every day.

Those strengths remain real.

The question is whether the country will organize itself well enough to use them wisely.

Renewal also requires discipline.

Strong nations do not improve through outrage alone. They improve through long-term thinking, competent leadership, responsible institutions, and citizens willing to invest effort into their communities and future.

The modern world often rewards distraction, short attention spans, and constant conflict. But rebuilding a nation requires patience. It requires maintaining systems before they collapse, educating children seriously, investing in infrastructure consistently, supporting productive work, and creating environments where people can build stable lives over time.

There are no shortcuts to lasting national strength.

America must also rebuild trust.

People need to believe that institutions still work, that effort still matters, and that the country is capable of solving problems instead of simply managing decline. Without trust, division grows deeper and public confidence weakens further.

Trust cannot be demanded.

It must be earned through competence, fairness, accountability, and visible results.

That may be one of the most important lessons of all.

Strong nations are not sustained by rhetoric alone. They are sustained when ordinary people can see real improvements in daily life:

- better schools,
- safer neighborhoods,
- more affordable living,
- stronger healthcare,
- stable jobs,
- modern infrastructure,
- and genuine opportunity.

Progress becomes believable when people can experience it directly.

Most importantly, America must recover a sense of shared responsibility.

A healthy nation cannot survive if every institution, every community, and every citizen acts entirely for themselves alone.

Renewal requires cooperation across families, schools, businesses, local governments, and communities. It requires recognizing that national strength is built collectively, even in a society that values individual freedom.

That balance matters.

America's greatest moments often came when ambition was matched with responsibility, innovation was matched with discipline, and freedom was matched with civic purpose.

The country does not need to become perfect to become stronger.

It simply needs to become serious again about the fundamentals that allow societies to thrive.

Because the future is still unwritten.

The systems that feel broken today can be repaired.
The communities struggling today can recover.
The confidence that has weakened can return.

History shows that nations rise when people choose to rebuild instead of surrendering to decline.

America still has that choice.

And the work of building a stronger nation begins now.

Afterword

Hope for the Future

America's future has not been decided yet.

Despite the challenges facing the country—rising costs, social division, addiction, declining trust, struggling institutions, and economic uncertainty—the story of America is still being written. History has shown many times that strong nations can recover when they are willing to confront problems honestly and rebuild with purpose.

That is the central message of this book:

Decline is not destiny.

No country remains strong automatically. Nations rise or fall based on the choices they make, the systems they build, and the willingness of their people to work toward something larger than themselves. America's greatest strength has never been perfection. Its greatest strength has been the ability to adapt, innovate, recover, and keep moving forward even during difficult times.

That strength still exists.

Across the country, millions of people continue to work hard, raise families, educate children, care for neighbors, build businesses, serve communities, and believe that tomorrow can still be better than today. Those quiet acts of responsibility and perseverance remain the foundation of national renewal.

This book has argued that America can rise again by strengthening the essentials:

- public health,
- education,
- housing,
- public safety,
- infrastructure,
- stable work,
- agriculture,
- innovation,
- and long-term competitiveness.

These are not abstract political ideas.

They are the building blocks of everyday life.

When people can afford healthcare, they live with greater stability.

When education is strong and affordable, opportunity expands.

When housing becomes reachable, families regain security.

When communities are safer, trust grows.

When infrastructure works, businesses and workers thrive.

When agriculture remains productive, the nation becomes more resilient.

When innovation is encouraged, economic strength returns.

Step by step, strength compounds.

That is how renewal happens.

Not through miracles.

Not through anger alone.

Not through endless division.

But through steady improvement, practical leadership, shared responsibility, and long-term thinking.

Hope is an important part of that process.

Real hope is not blind optimism or denial of problems. Real hope is the belief that problems can still be solved through effort, discipline, and cooperation. It is a refusal to accept decline as permanent. It is the decision to keep building even when rebuilding is difficult.

America does not need to become a perfect nation to become a stronger one.

It only needs to recover confidence in its ability to do the basics well again:

- educate children seriously,
- support families,
- reward hard work,
- maintain infrastructure,
- protect public safety,
- encourage innovation,

and create conditions where ordinary people can build meaningful lives.

Strong nations are not built in a single moment.

They are built generation after generation by people willing to invest in the future instead of surrendering to cynicism.

The future of America will depend on whether the country chooses renewal over decline, discipline over complacency, and action over despair.

That choice still exists.

And as long as that choice exists, there is reason for hope.

Because America has risen before.

And it can rise again.

About the Author

Jean R. Jean Philippe is a Homeland Security graduate from Keiser University with a strong commitment to strengthening America's resilience, prosperity, and long-term stability. With a background rooted in national security principles and a passion for community well-being, Jean R. Jean Philippe brings a clear, solutions-focused perspective on the challenges facing the nation.

Driven by the belief that a stronger America begins with informed citizens and practical ideas, Jean writes to bridge the gap between everyday people and the policies that shape their lives. *Building a Stronger America: A Roadmap to Health, Prosperity, and National Renewal* is his debut work, offering a grounded and accessible vision for a healthier, more secure, and more unified nation.

Jean Philippe lives in Florida, where he continues to study the evolving landscape of national security, public policy, and America's future.

www.ingramcontent.com/pod-product-compliance
Lightning Source LLC
LaVergne TN
LVHW050939210926
848482LV00013B/672
9798234114433